discovermore
The Human Body

How Our Skeletal System Works

Dwight Morris

IN ASSOCIATION WITH

Published in 2026 by Britannica Educational Publishing (a trademark of Encyclopædia Britannica, Inc.) in association with The Rosen Publishing Group, Inc.
2544 Clinton Street, Buffalo, NY 14224

Distributed exclusively by Rosen Publishing.
To see additional Britannica Educational Publishing titles, go to rosenpublishing.com.

Portions of this work were originally authored by Laura Loria and published as *The Bones in Your Body*. All new material in this edition was authored by Dwight Morris.

Editor: Greg Roza
Book Design: Rachel Rising

Photo Credits: Cover, (series background) Dai Yim/Shutterstock.com; Cover, Sebastian Kaulitzki/Shutterstock.com; p. 4 Philipp Nicolai/Shutterstock.com; p. 5 Twinkle picture/Shutterstock.com; p. 7 DestinaDesign/Shutterstock.com; p. 7 LukaSkywalker/Shutterstock.com; p. 8 Adisak Riwkratok/Shutterstock.com; p. 9 stihii/Shutterstock.com; p. 10 stihii/Shutterstock.com; p. 11 sciencepics/Shutterstock.com; p. 12 4zevar/Shutterstock.com; p. 13 Aurelie Fieschi/Shutterstock.com; p. 15 Alila Medical Media/Shutterstock.com; p. 15 Dikushin Dmitry/Shutterstock.com; p. 16 Rawpixel.com/Shutterstock.com; p. 17 Vecton/Shutterstock.com; p. 19 Lucky Business/Shutterstock.com; p. 19 Anna Puzatykh/Shutterstock.com; p. 20 Aleksandar Malivuk/Shutterstock.com; p. 21 Peter Porrini/Shutterstock.com; p. 22 Designua/Shutterstock.com; p. 23 Ayah Raushan/Shutterstock.com; p. 24 Brian Glowacki/Shutterstock.com; p. 25 Pikovit/Shutterstock.com; p. 26 ChooChin/Shutterstock.com; p. 27 Terelyuk/Shutterstock.com; p. 28 Denis---S/Shutterstock.com; p. 29 MarinaGrigorivna/Shutterstock.com.

Cataloging-in-Publication Data
Names: Morris, Dwight.
Title: How our skeletal system works / Dwight Morris.
Description: New York : Britannica Educational Publishing, in association with Rosen Educational Services, 2026. | Series: Discover more: the human body | Includes glossary and index.
Identifiers: ISBN 9781641904490 (library bound) | ISBN 9781641904483 (pbk) | ISBN 9781641904506 (ebook)
Subjects: LCSH: Human skeleton--Juvenile literature. | Bones--Juvenile literature. | Human physiology--Juvenile literature.
Classification: LCC QM101.M677 2026 | DDC 611'.71--dc23

Manufactured in the United States of America

Some of the images in this book illustrate individuals who are models. The depictions do not imply actual situations or events.

CPSIA Compliance Information: Batch #CSBRIT26. For further information contact Rosen Publishing at 1-800-237-9932.

Contents

The Human Skeleton

Bones are your body's hidden frame. Bones provide a structure, called the skeletal system, for your other body parts. They are connected to your muscles, and they make it possible for you to walk and run. Some bones protect the organs inside your body, such as your lungs and brain. The smallest bones in the body allow us to hear.

Our bones wouldn't be very useful if we didn't have muscles too! Bones and muscles work together to move our body.

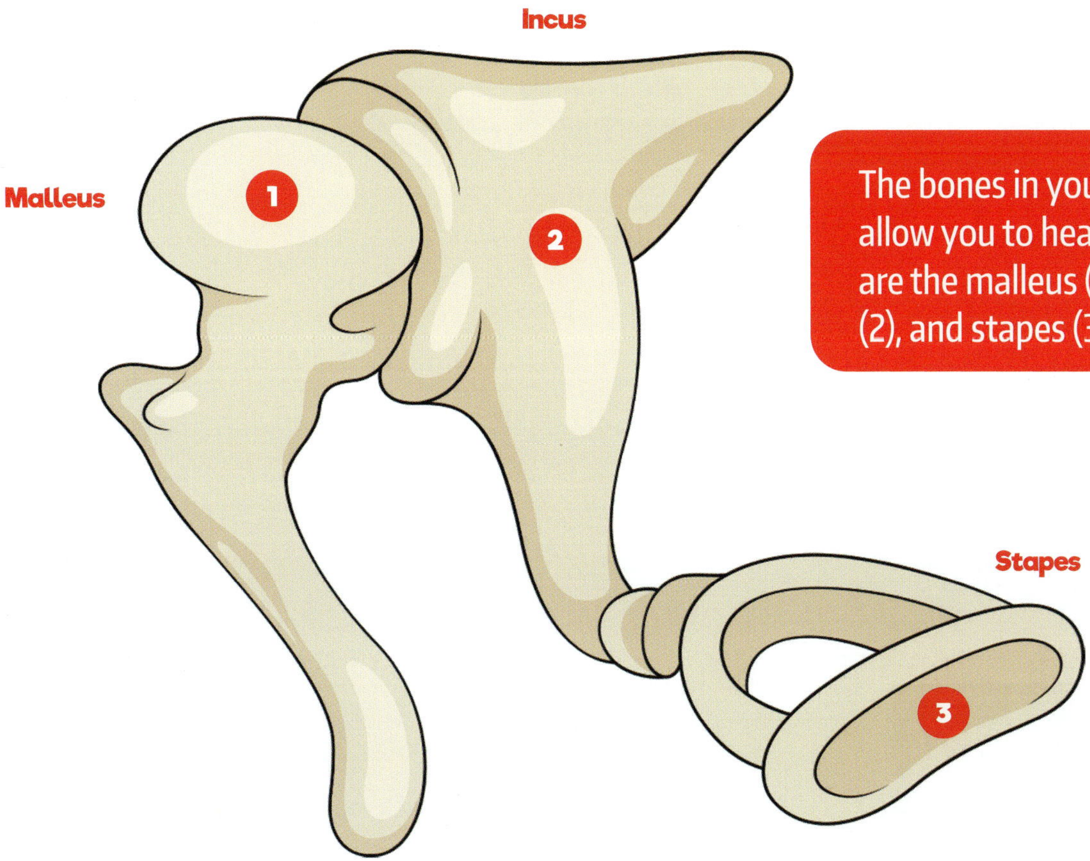

Incus

Malleus

Stapes

The bones in your ear allow you to hear. They are the malleus (1), incus (2), and stapes (3).

From the outside, a bone looks hard and still. But inside, it is always working. Bones store nutrients, or things your body needs. Bones also make blood cells. The area inside your bones is filled with a soft matter called marrow.

Consider This

How would your life be different if you didn't have bones?

Inside Bones

If you look at a bone that has been cut, you might be surprised to see that it has several layers. The outer layer of a bone is compact bone. It is firm and dense. The inner layer is spongy bone. It looks like a sponge, with many holes, but it is not soft or flexible. The holes in spongy bone are filled with bone marrow. Marrow is soft like jelly and can be red or yellow. Most blood cells are created by red marrow, while fat is stored in yellow marrow.

All bones are covered with a membrane, or skin. The membrane contains nerves and blood vessels that are attached to the bone. They carry blood cells from the bone to the rest of the body. They also carry nutrients to the bone.

Bones are made of calcium and other minerals, protein fibers, and water.

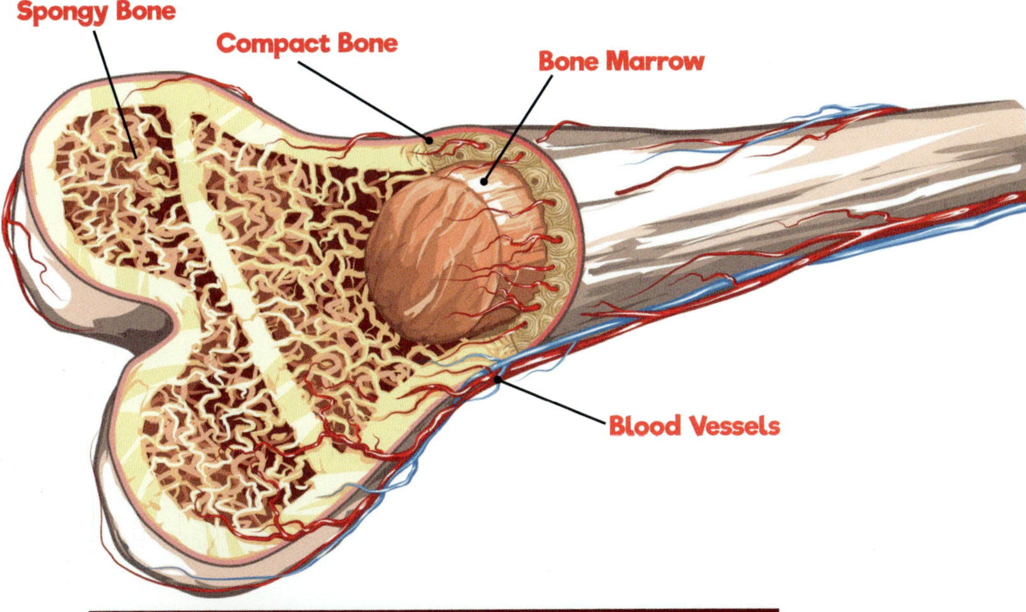

compareandcontrast

How is compact bone different from the spongy bone inside it? And how is marrow different from bone?

This section of a bone shows both compact bone and spongy bone.

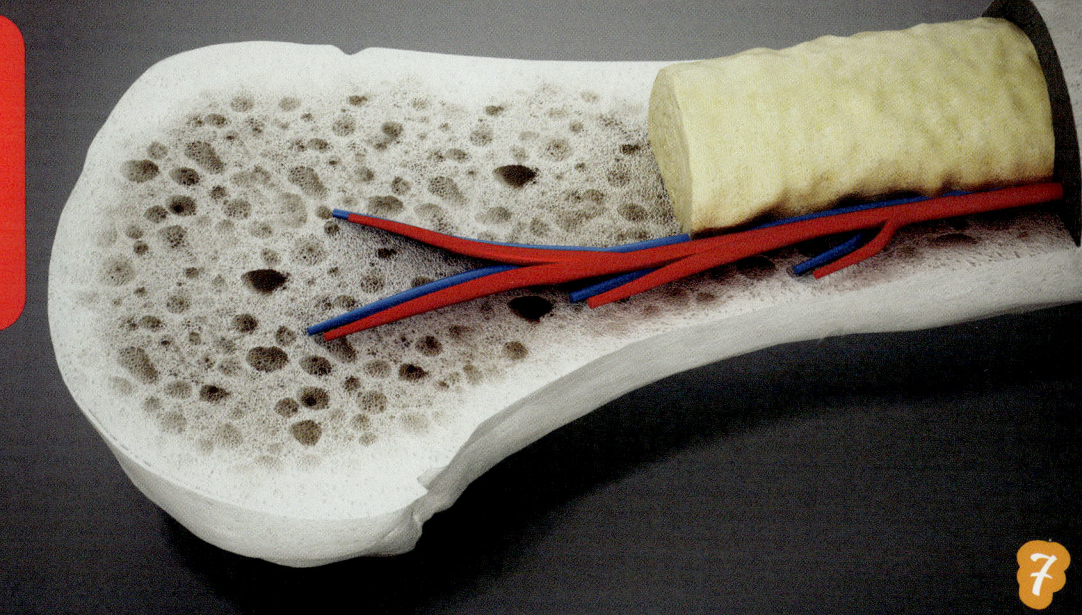

Get Connected

The skeletal system is made up of all of the body's bones and the **tissues** that connect them. Bones are held together by strips of tissue called ligaments. In between the bones is a cushion made of cartilage, which is strong but flexible. Cartilage keeps bones from grinding against each other. Joints are structures that hold together two or more bones. Joints give the bones support and allow the skeleton to move.

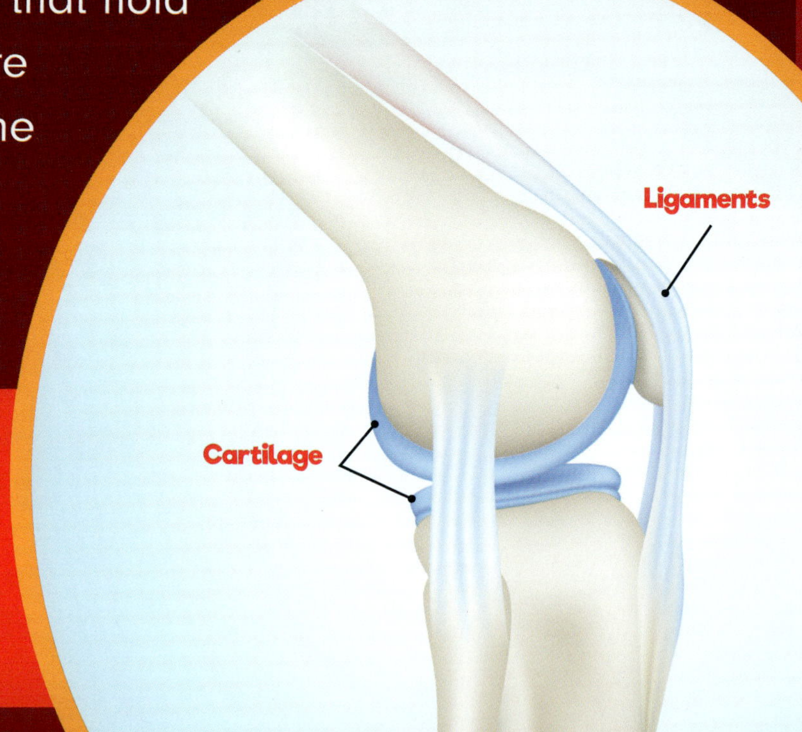

Cartilage helps protect bones and joints. Another kind of cartilage makes up the nose and ears.

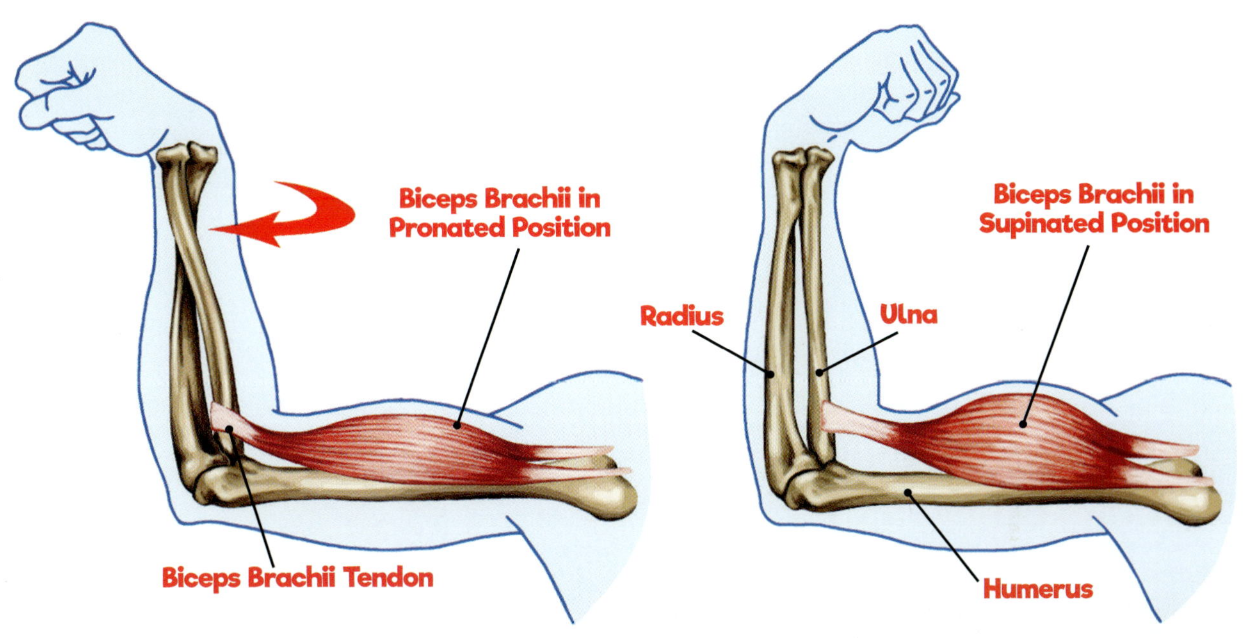

Many of the bones in your body require one set of muscles to move them one way and another set of muscles to move them back. This diagram shows how the muscles and bones of the arm work.

Bones are connected to muscles by tendons, another kind of tough cord. Bones work with muscles to make the body move. In order to move, the brain sends a signal to the muscles. The muscles then pull or push on the bone.

WORD WISE

TISSUES ARE GROUPS OR LAYERS OF CELLS THAT WORK TOGETHER TO PERFORM A SPECIFIC JOB.

Bone Types

The human body contains four types of bones. Leg and arm bones are examples of long bones. They look like tubes with two bumps at each end. Long bones help the body move. Short bones are found in the ankles and wrists. They are cube-shaped and consist mostly of spongy bone.

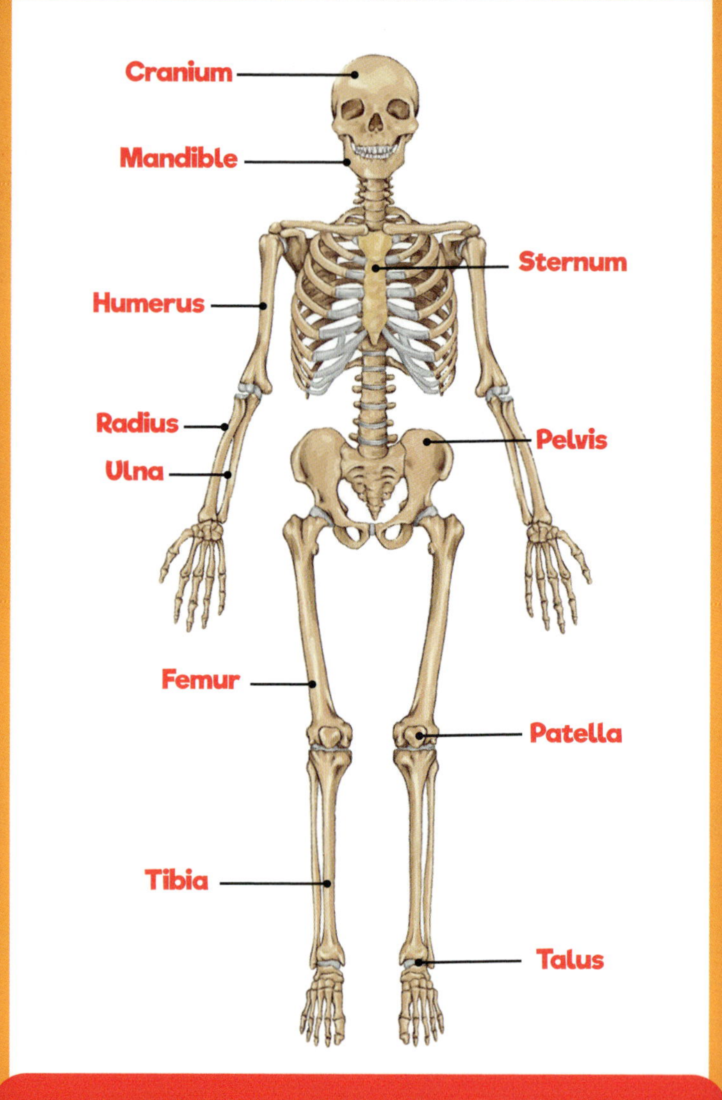

The adult human body has 206 bones.

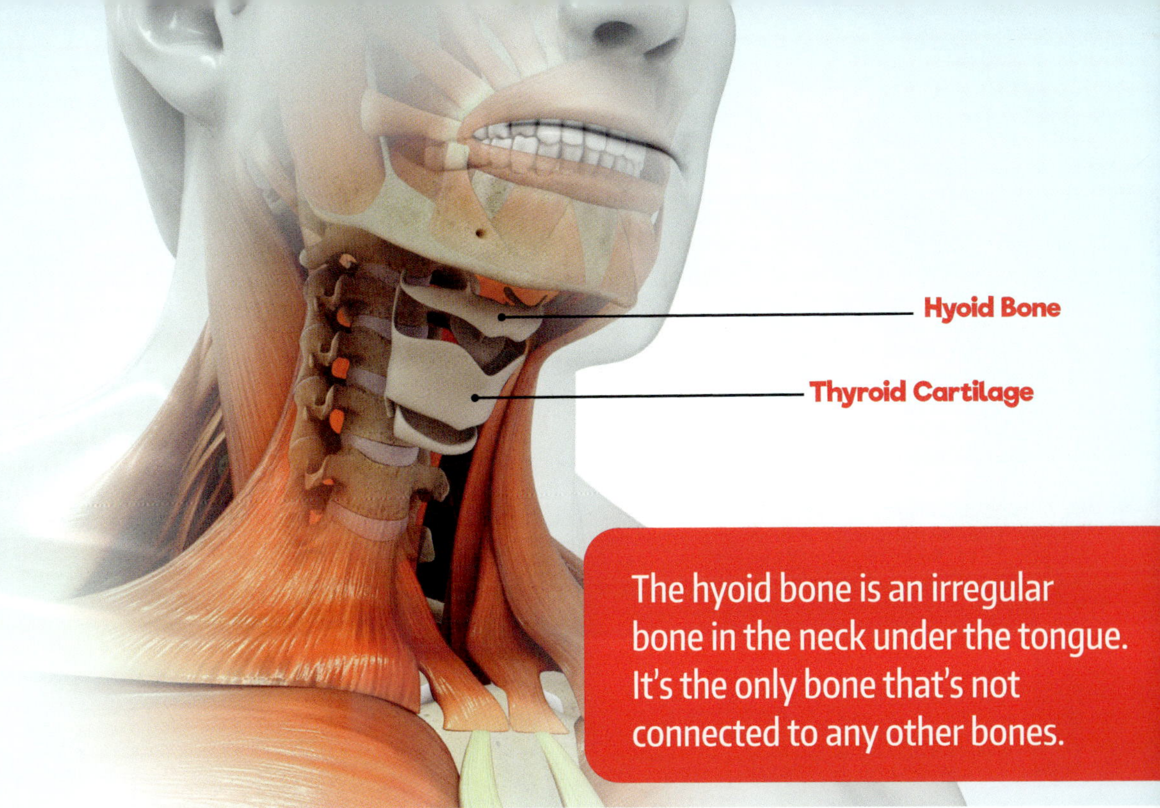

The hyoid bone is an irregular bone in the neck under the tongue. It's the only bone that's not connected to any other bones.

The bones in the skull and rib cage are flat bones. They are often thinner than other bones. They protect important organs from damage. Irregular bones don't have any particular shape. They are each unique. Irregular bones include the bones in the backbone and the upper and lower jawbones.

Consider This

Where on you body can you feel your bones? Can you detect any of the traits mentioned here?

The Backbone

Your backbone, or spine, isn't just a single bone. It's made up of many smaller bones. The spine is easy to find. It runs up and down your back. You can feel its bumpy shape through your skin. The spine helps you stand upright.

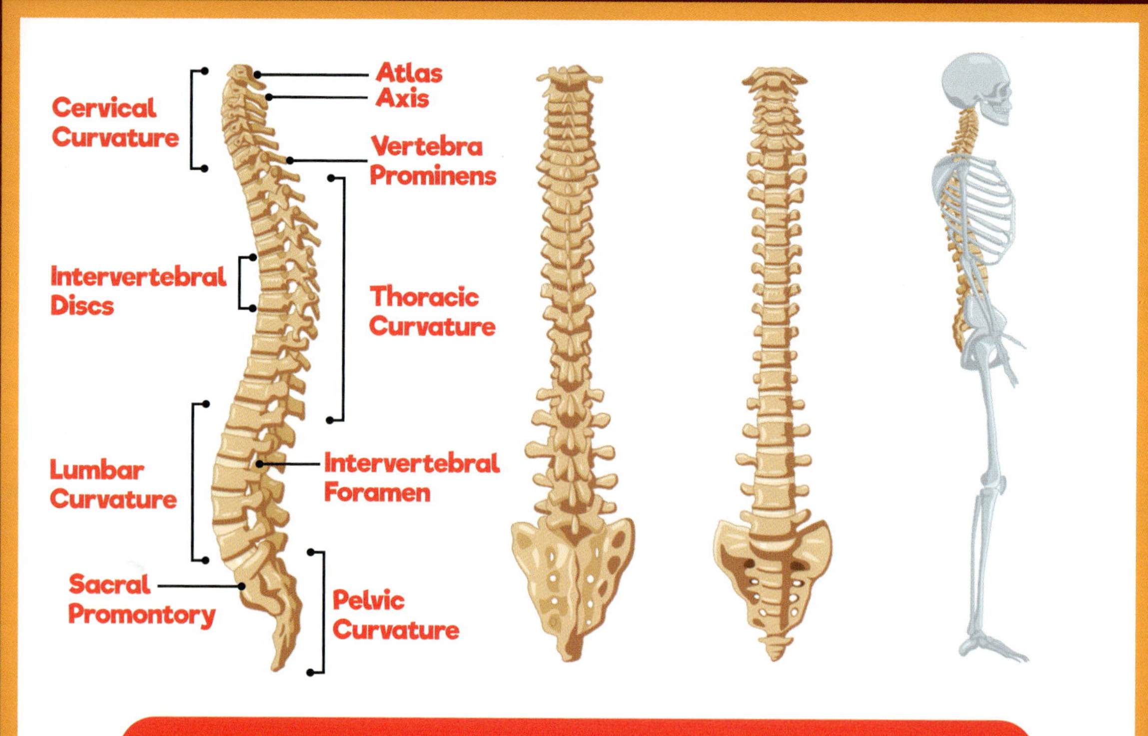

Vertebrae are the bones that make up the spine. A single one of these bones is called a vertebra.

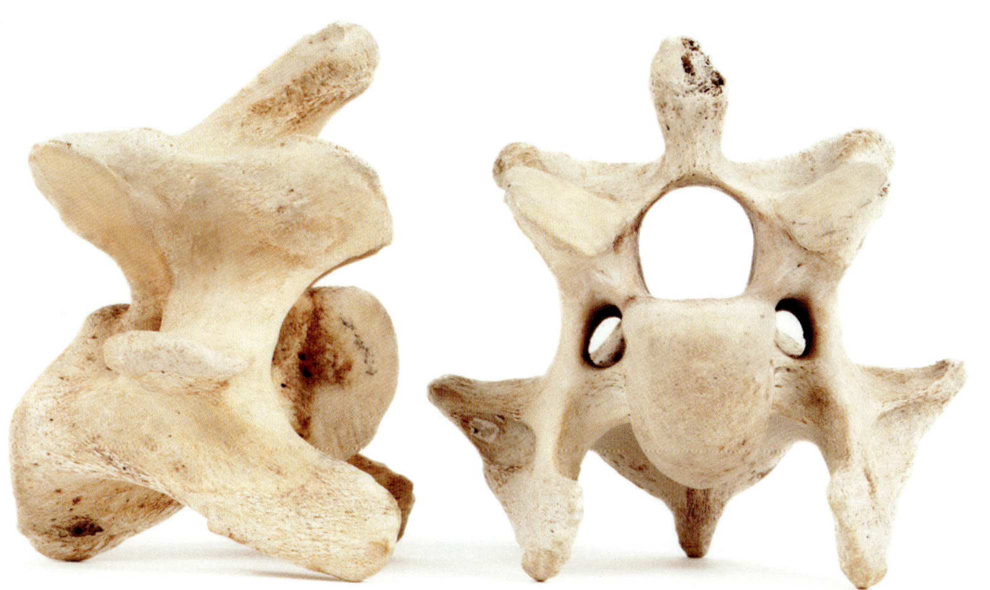

Each vertebra has a hole in the middle. The spinal cord, a thick bundle of nerves, runs through these holes. The spine protects the spinal cord.

The backbone is made up of 33 vertebrae. Most of the vertebrae are ring shaped. Between most of them is a disk of cartilage. There are several groups of vertebrae that do different jobs. At the top of the spine, the cervical vertebrae support the head. Underneath that, the thoracic vertebrae are attached to the ribs, which protect the heart and lungs. Lumbar vertebrae toward the bottom of the spine provide strength and balance.

Consider This

Why do you think the spine is made up of many smaller bones? What would happen if the spine was one long bone?

The Bones of the Hand

The bones in the hand are connected in ways that let you grip, push, and pull objects. A complex system of bones, cartilage, tendons, and muscles allows our hands to do a wide range of tasks.

There are 27 bones in each hand. Each finger has three, and the thumb has two. This group is called the phalanges. The palm has five bones. They are the metacarpals. All of these are long bones, which means that they are longer than they are wide. They are connected by joints.

In the wrist, there are eight short bones. These are called carpal bones. Along with four joints, the carpal bones allow your hand to bend and twist. Each small bone of the hand has its own name.

The hand is made up of long bones and short bones.

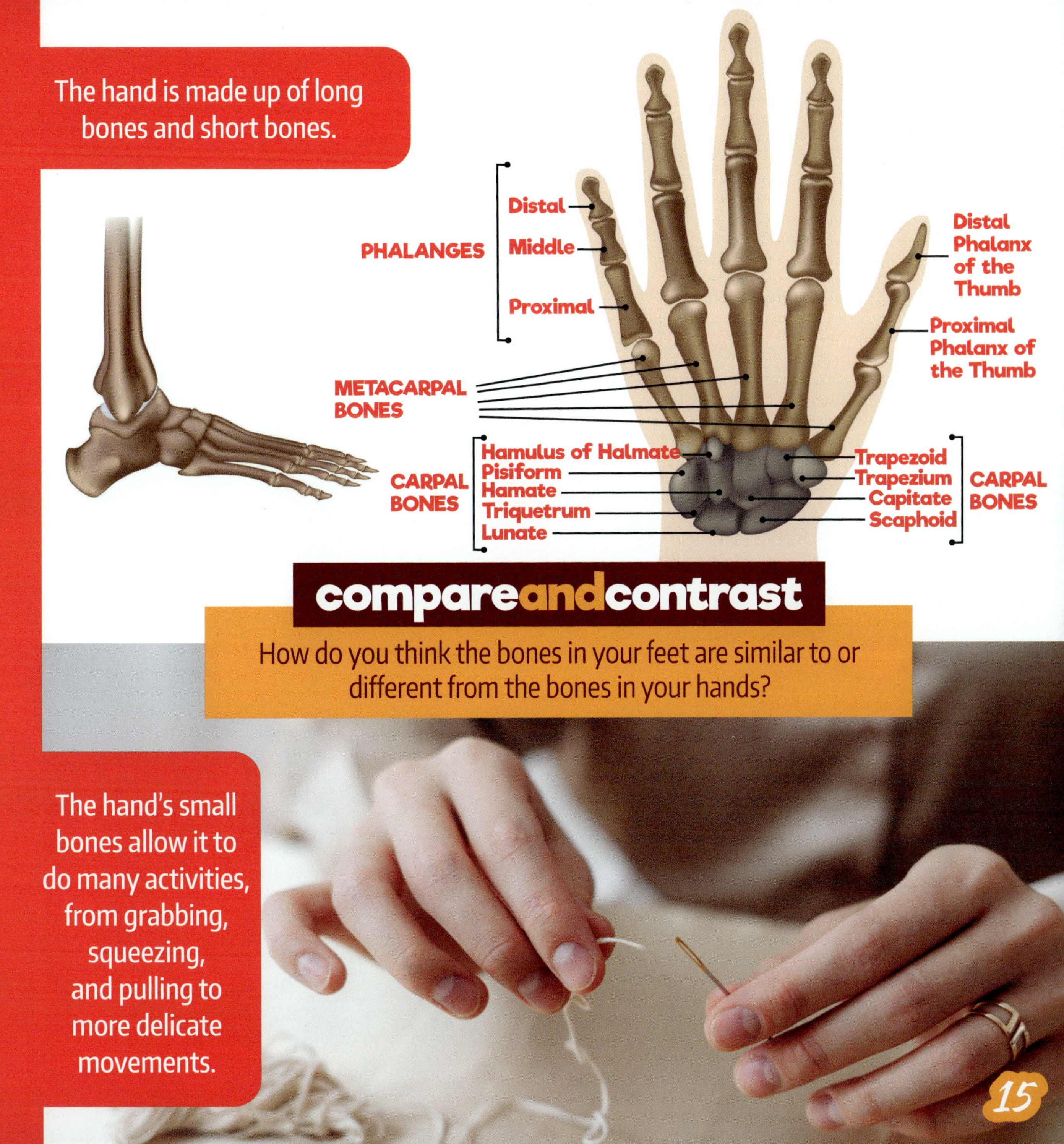

compareandcontrast

How do you think the bones in your feet are similar to or different from the bones in your hands?

The hand's small bones allow it to do many activities, from grabbing, squeezing, and pulling to more delicate movements.

The Skull

The skull is made up of hard, flat bones. The skull acts like a shell for your brain and protects it from injury. It also gives your face its shape.

The cranium, on the top and back of your head, covers your brain. It feels like one piece, but it is actually made up of several bones that are joined together. A baby's cranium has spaces between these bones. As a baby grows, his or her bones meet up at joints that do not move.

The shape of the skull determines what your face looks like.

The bones of the skull are connected with a fibrous tissue called sutural ligament. They form a joint called a fibrous joint. "Fibrous" means resembling fibers, or strands.

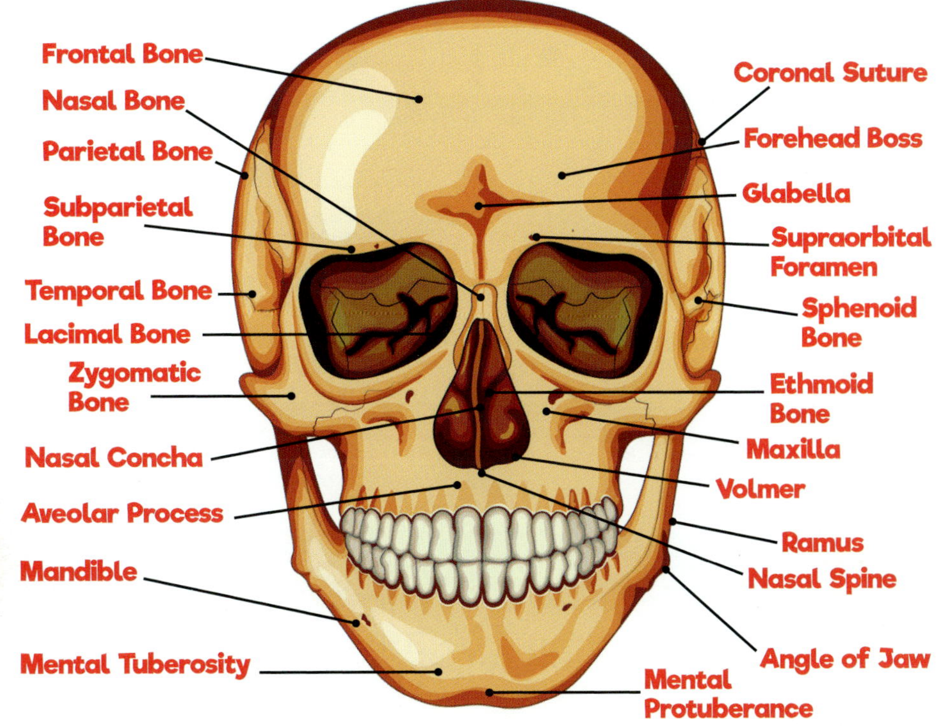

The rest of the skull makes up your face. There are holes that support and protect your eyes and an opening that allows air to come through your nose. There is also a hole for the spinal cord to connect to your brain.

Consider This

The bones of the skull are very strong but can break, or fracture. How can you protect your skull from fractures and other injuries?

Bone Health

How do we keep our bones healthy? First, it's important to eat foods that have the vitamins and minerals that help build bone. Those foods include milk and other dairy products, leafy green vegetables such as spinach and kale, and some fish.

Exercise also helps make your bones strong. It is good for your muscles and your overall health as well. However, it is important to protect your bones while you are taking part in physical activities. To give your bones even more protection, be sure to wear protective gear like helmets or shin guards while playing sports.

Having several servings of calcium-rich foods every day is good for bone health.

compare and contrast

Think about the types of protective gear a hockey player wears compared to the gear a tennis player wears. Why are they different?

Hockey is a hard-hitting game! Players protect their bones by wearing special gear, including helmets, facemasks, chest protectors, and heavy gloves.

A Look Inside the Body

Sometimes a doctor needs to see a patient's bones. The most common way for a doctor to view bones is to use an X-ray machine. X-rays are powerful waves of energy. These rays can go through some material that light cannot go through. To see the bones inside a body, the rays are sent through the body. The X-rays pass through the skin and soft tissues, such as muscle. But hard body parts, such as bones, block the X-rays. Special film captures the image made by the X-rays.

Someone trained to use X-ray technology to take X-rays is called a radiologic technologist.

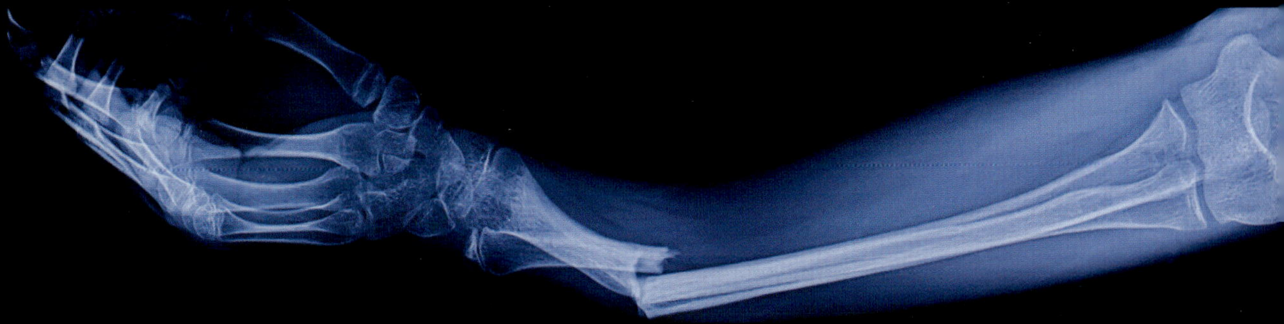

This in an X-ray of a forearm with two broken long bones—the ulna and the radius.

Sometimes, an X-ray picture is not clear enough. In that case, a doctor might order other types of tests that give a clearer image. Although these tests usually take longer, they give a doctor more information, allowing them to make a more accurate diagnosis.

Consider This

Can you think of other uses for an X-ray machine other than seeing our bones?

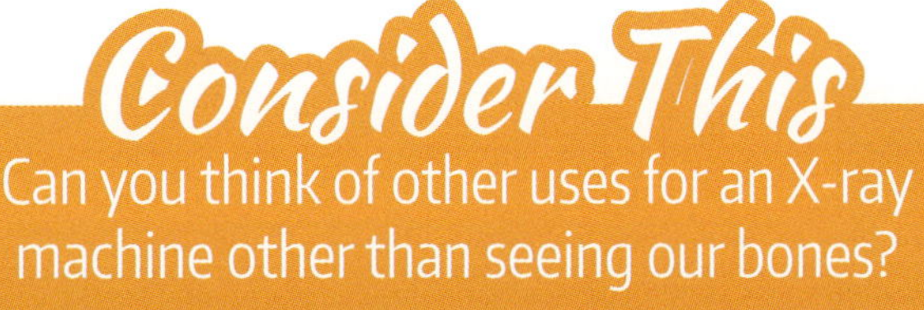

Bone Breaks

Fractures, or broken bones, happen when too much pressure is placed on a bone. When a bone breaks all the way through but stays under the skin, it is a simple fracture. When a bone breaks and pokes through the skin, it is called a compound fracture.

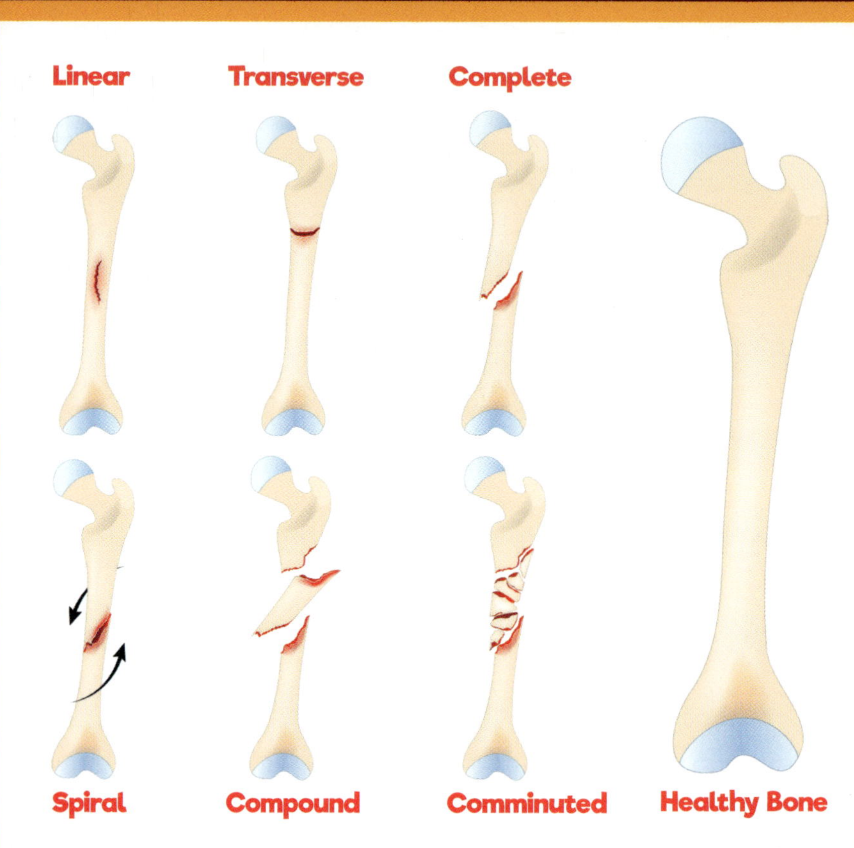

A compound fracture is more dangerous because it creates an open wound.

It can be hard to get around with a heavy cast on your leg! However, newer cast materials are much lighter and more comfortable.

Doctors use X-rays to see what kind of fracture a patient has. The doctor then sets the bone back in place, sometimes using metal pins, and usually puts a cast over it. A cast can be made of plaster, fiberglass, or plastic. A cast protects the broken bone and keeps it still so it can heal.

Consider This

Children that break a bone often suffer a greenstick fracture. This happens when a bone bends and cracks on one side only. Why do you think greenstick fractures are far more common in kids than in adults?

Diseases of the Bones

Most bone diseases affect older adults. However, practicing healthy habits when you are young can prevent you from developing some bone diseases.

Arthritis is a disease of the joints, where bones meet. It causes painful swelling. People with arthritis in their hands and wrists have a hard time holding or squeezing things. Arthritis in the knees makes it hard to walk.

Severe arthritis in old age can cause heavy swelling and misshapen fingers.

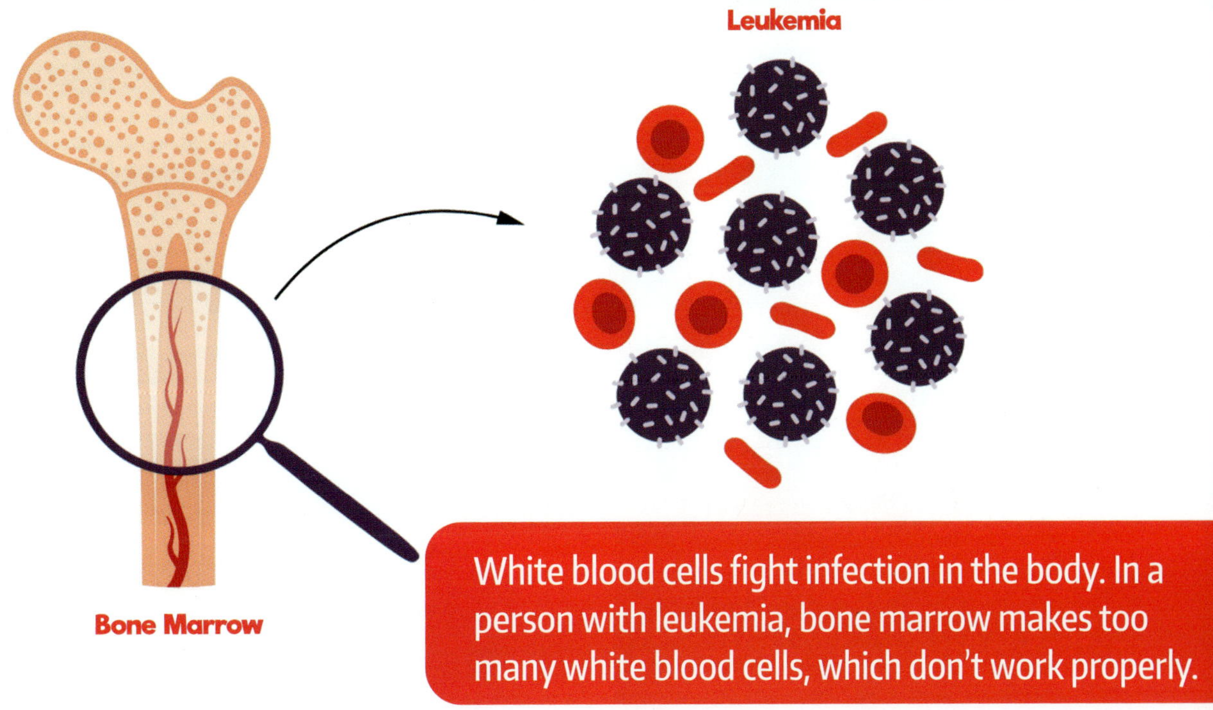

White blood cells fight infection in the body. In a person with leukemia, bone marrow makes too many white blood cells, which don't work properly.

After about age 40, people begin to lose some of the material in their bones. If they lose too much, their bones may break easily. Low bone **density** is sometimes caused by poor diet and lack of exercise.

Leukemia is a blood cancer that can occur in the bone marrow. The white blood cells of people with leukemia do not work correctly. This makes it hard for their bodies to fight infections.

WORD WISE

THE WORD DENSITY REFERS TO HOW TIGHTLY PACKED MATTER IS.

Replacing Bones

Bones can sometimes heal themselves by creating new tissue. Special cells start making new bone by building a structure of protein. Blood brings in calcium, which sticks to the protein structure. The calcium builds up and hardens, creating the bone's structure. Sometimes, however, bones cannot heal themselves. Joints can wear out over time, or disease can wear away at bones. Scientists therefore have come up with several ways to replace bones.

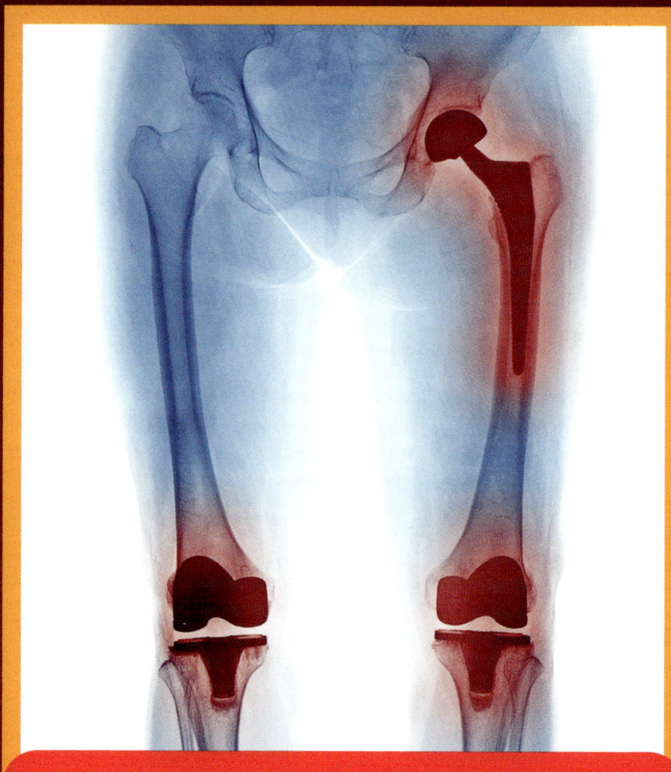

This X-ray shows that this person has had two knee replacements and one hip replacement.

This doctor is performing a bone marrow transplant.

Surgeons can replace hips, knees, shoulders, and many other joints and bones. The replacements are usually made of metals, ceramics, and plastics.

Bone marrow creates blood cells for our bodies, so it is important for it to stay healthy. Doctors may need to perform a bone marrow transplant if it becomes diseased.

Consider This

When a donor is needed for a bone marrow transplant, doctors first check with family members. Why would family members make the best bone marrow donors?

In the future, doctors may be able to replace whole bones with **artificial** bones. Rattan, a type of wood, has been used to replace bones in animals after it has been treated with chemicals and heated. Other scientists are using forms of plastic to create molds for metal replacement bones.

Doctors use the metal titanium for artificial joints because it's light, the body doesn't reject it, and it doesn't cause illnesses.

This artificial titanium jawbone was created with a 3-D printer.

Another exciting new tool for doctors is 3-D printing. A 3-D printer works like a computer printer, but instead of printing a flat page, it uses plastics and other substances to create a solid object from exact measurements. The object can be placed in the body, where bone cells can attach to it and grow new bone. The idea is to make a bone that is a perfect match for the old bone that the body won't reject.

WORD WISE

ARTIFICIAL MEANS MANMADE; NOT NATURAL.

Glossary

bone marrow A soft substance that fills the bones of people and animals.

cancer A serious disease caused by cells that are not normal and that can spread to one or many parts of the body.

cartilage Strong, flexible material found between bones.

cells The smallest units of living matter that can exist by themselves.

complex Made of many parts.

diagnosis To recognize an illness by its symptoms, or signs.

injury Harm or damage.

irregular Not having a standard size or shape.

joints Places where bones meet.

ligament Tissue that holds bones in place, like a rubber band.

membrane A thin, protective cover.

minerals Substances (such as iron or zinc) that occur naturally in certain foods and are important for good health.

muscles Body tissue that can contract and produce movement.

nutrients Natural substances that the body needs to function.

organs Body parts that have a particular job.

protein A substance found in foods (such as meat, milk, eggs, and beans) that is an important part of the human diet.

tendons Bands that connect muscles to bones.

transplant Taking something from one place and putting it in another place.

For More Information

Books

Faust, D. R. *The Skeletal System.* Minneapolis MN: Bearport Publishing, 2025.

Xiong, Keng. *What Is the Skeletal System?* Parker, CO: The Child's World, 2025.

Websites

Bones and the Human Skeleton
www.ducksters.com/science/bones.php
Learn more about the skeletal system. Includes helpful links to related topics.

Your Bones
kidshealth.org/en/kids/bones.html
Read even more about your bones!

Publisher's note to educators and parents: Our editors have carefully reviewed these websites to ensure that they are suitable for students. Many websites change frequently, however, and we cannot guarantee that a site's future contents will continue to meet our high standards of quality and educational value. Be advised that students should be closely supervised whenever they access the internet.

Index